AF521906

Dangerous Bugs

KISSING BUGS

MEGAN COOLEY PETERSON

BLACK RABBIT BOOKS

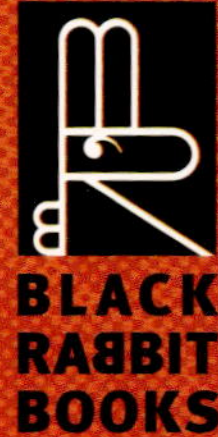

BOLT

Bolt is published by Black Rabbit Books
P.O. Box 227, Mankato, Minnesota, 56002
www.blackrabbitbooks.com

Alissa Thielges, editor; Michael Sellner, designer
and photo researcher

Library of Congress Cataloging-in-Publication Data
Names: Peterson, Megan Cooley, author.
Title: Kissing bugs / by Megan Cooley Peterson.
Description: Mankato, Minnesota: Black Rabbit Books, [2024] |
Series: Bolt: Dangerous bugs | Includes bibliographical references and index. |
Audience: Ages 8–12 | Audience: Grades 4–6 |
Summary: "Able to deliver a deadly disease, kissing bugs are more than just a pest. Get up close to these dangerous bugs through gross photos, leveled text, and engaging infographics that'll make readers squirm"—Provided by publisher.
Identifiers: LCCN 2022024135 (print) | LCCN 2022024136 (ebook) |
ISBN 9781623105792 (library binding) | ISBN 9781623105853 (ebook)
Subjects: LCSH: Conenoses—Juvenile literature.
Classification: LCC QL523.R4 P48 2024 (print) | LCC QL523.R4 (ebook) |
DDC 595.7/54—dc23/eng/20220624
LC record available at https://lccn.loc.gov/2022024135
LC ebook record available at https://lccn.loc.gov/2022024136

Printed in China

Image Credits

Alamy: Evgeny Epifantsev 6, Nigel Cattlin 16–17, 20, Ray Wilson 6, 8–9, 20, 22–23, Tody Kuswardioko 15; Flickr: Dr. Alexey Yakovlev 26–27, ZEISS Microscopy 25; iStock: Md Saiful Islam Khan 28–29; Science Source: David Scharf cover, 4–5, 28–29, Eye of Science 8–9, 22–23, 32, KATERYNA KON 25, Martin Dohrn 6, The Natural History Museum, London 1, Volker Steger 10; Shutterstock: casejustin 7, schlyx 12–13, schlyx 3, 31, ShutterstockProfessional 8–9, TRR 19; CDC: U.S. Department of Health & Human Services 18–19

CHAPTER 1

Sweet DREAMS?

It's the middle of a summer night. A kissing bug slips through a gap in a window. It enters a dark bedroom where a person sleeps quietly. The bug crawls onto the bed. It moves onto the person's face. The kissing bug is about to eat.

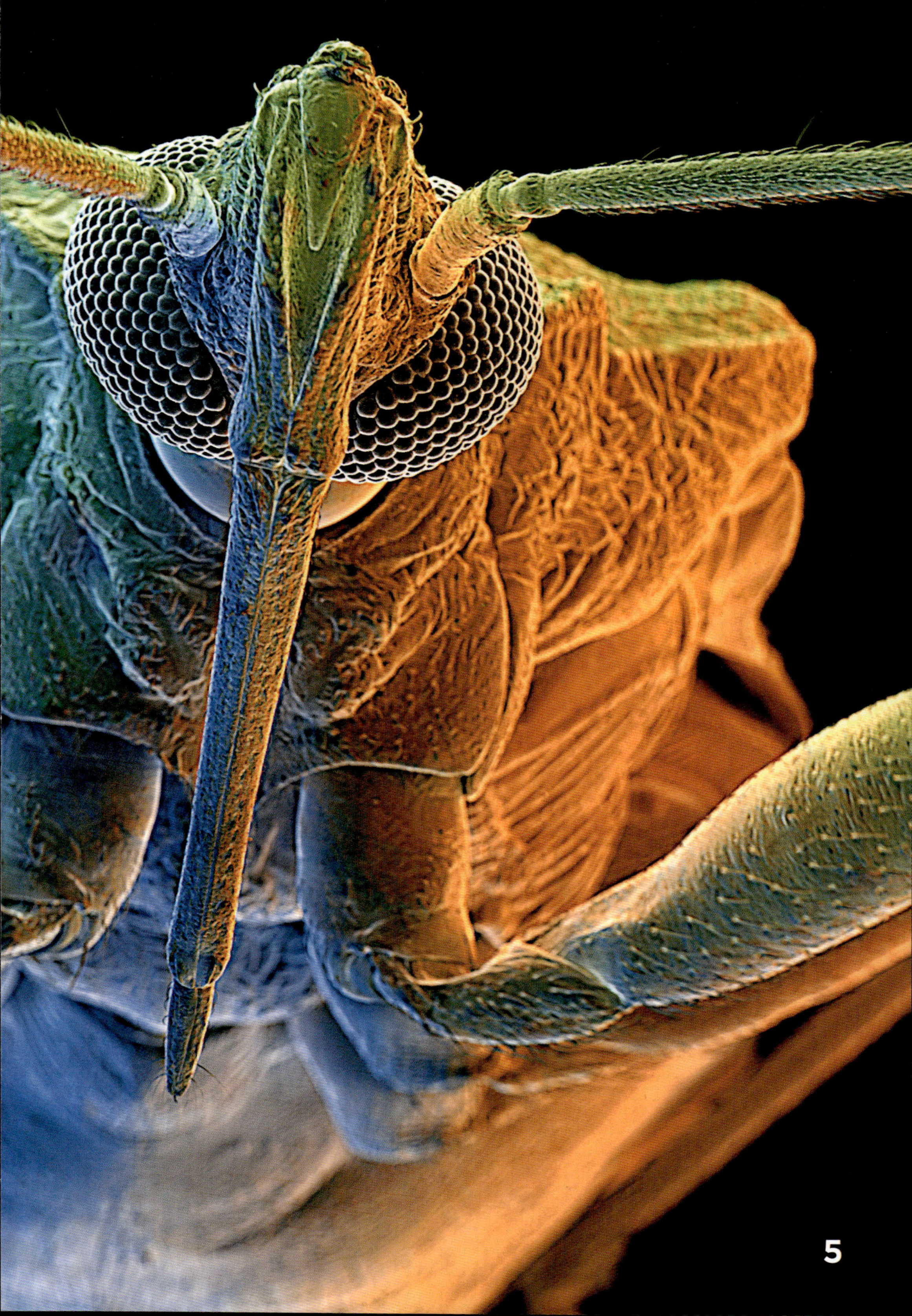

Feeding Time

In a quick movement, the kissing bug's mouthpart **pierces** the person's skin. It sucks up a blood meal. Then the kissing bug slinks away. It hides in a dark corner of the room. It will feed again soon.

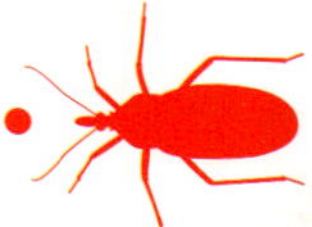

Kissing bugs usually bite people near the mouth. That's how they got their name.

CHAPTER 2

Size and

Kissing bugs are anything but sweet. These small insects look like beetles. They are about the size of a penny. Kissing bugs walk on six legs. Their bodies are mostly brown or black. They have orange or reddish markings on their **abdomens**.

How Long Is a Kissing Bug?

LIBERTY
LENGTH
0.5 to 1.25
INCH
(13 to 32 millimeter)

Bug Bodies

Kissing bugs have cone-shaped heads. They don't have chewing mouthparts like some bugs. Instead, kissing bugs have a **proboscis**. This mouthpart is tucked under their heads. When kissing bugs feed, they jab their mouthpart into prey. Then they suck up the blood.

Kissing bugs belong to a group of bugs called assassin bugs. These bugs often suck bodily fluids from people and animals.

PARTS OF A KISSING BUG

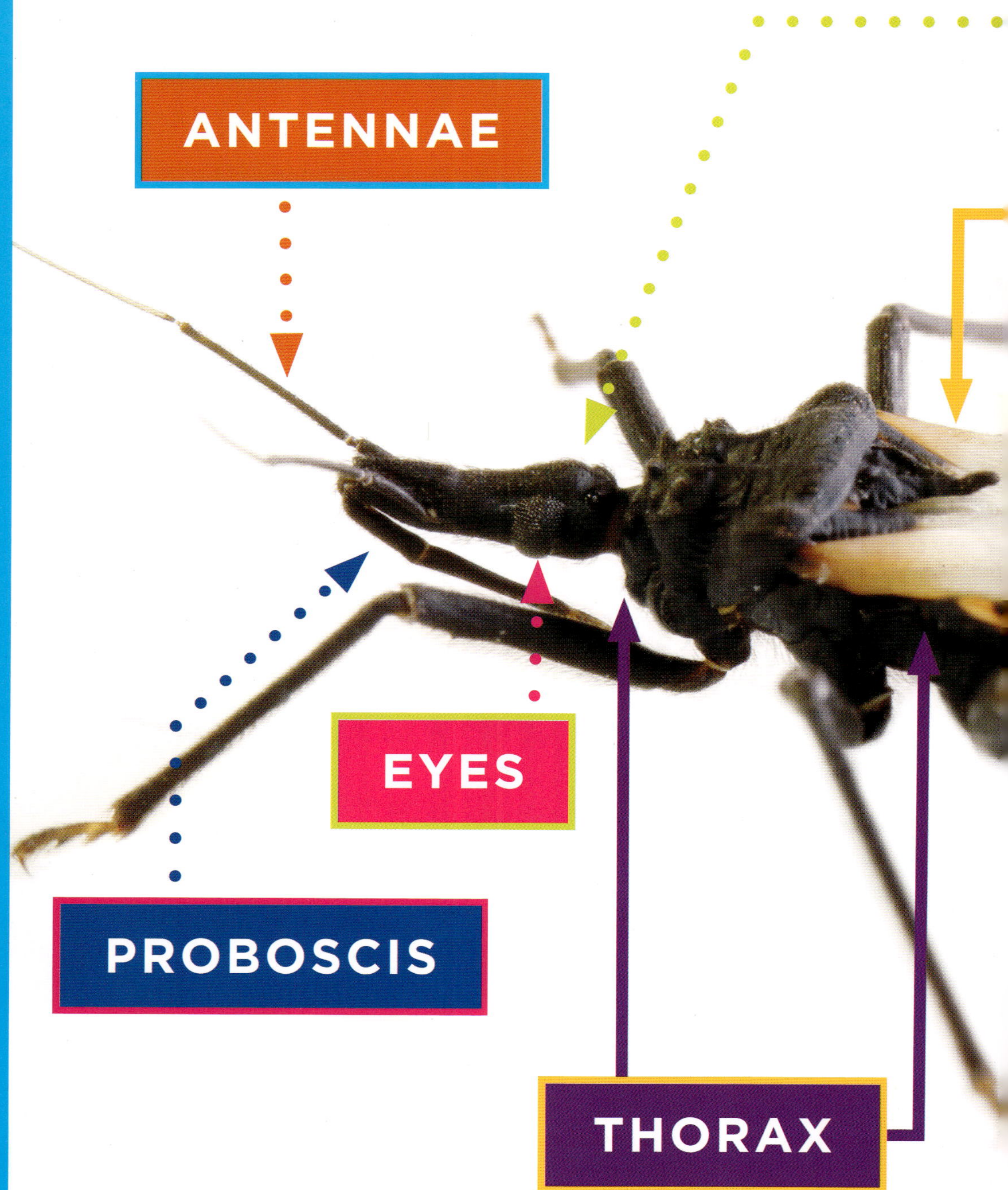

HEAD
ABDOMEN
WINGS
LEGS

CHAPTER 3

Where They Live and

Kissing bugs mostly live in Central and South America. They are also found in the southern United States. Kissing bugs like dark places. They hide out under porches or in piles of leaves. They live in soil and animal **burrows**. They even make homes under cement or in ceiling cracks.

Where Kissing Bugs Live

North America
Europe
Asia
Africa
South America
Australia

On the Hunt

When the sun goes down, kissing bugs go on the hunt. They feed on the blood of people and animals. Young kissing bugs have big **appetites** for blood. Some eat up to eight or nine times their body weight.

CHAPTER 4

Female kissing bugs lay small, white or pink eggs. Before laying eggs, they must eat blood. Kissing bugs only lay a few eggs at a time. These eggs hatch in about 10 to 40 days. They hatch faster in warm weather. Kissing bugs live and grow on their own.

Female kissing bugs can lay hundreds of eggs in a lifetime.

nymph

The Young

Young kissing bugs are called nymphs. They look like small adults. Nymphs go through many stages of growth. They molt between each stage. When nymphs molt, they shed their old **exoskeletons**. A new exoskeleton has grown underneath. Nymphs must eat blood before each molt.

Stages

From Egg to ADULT

Female kissing bugs lay eggs in warmer months.

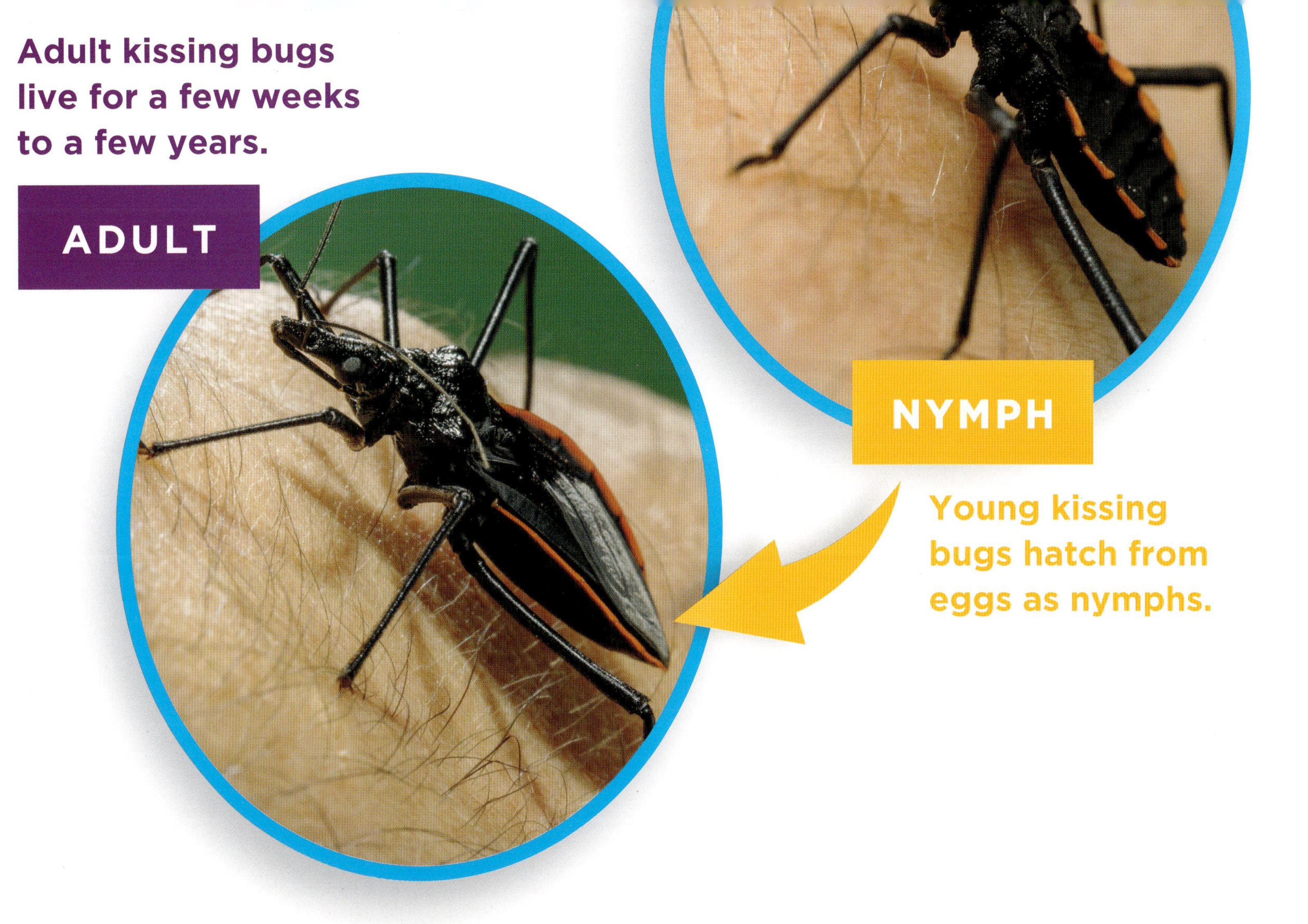
NYMPH
Young kissing bugs hatch from eggs as nymphs.
ADULT
Adult kissing bugs live for a few weeks to a few years.

CHAPTER 5

Out!

A kissing bug's bite usually doesn't hurt. But many kissing bugs carry a **parasite**. After a kissing bug feeds, it poops. When their waste enters a fresh bite, it can spread the parasite. The parasite can cause Chagas disease in people. Chagas disease causes fever and swelling. It can also damage the heart.

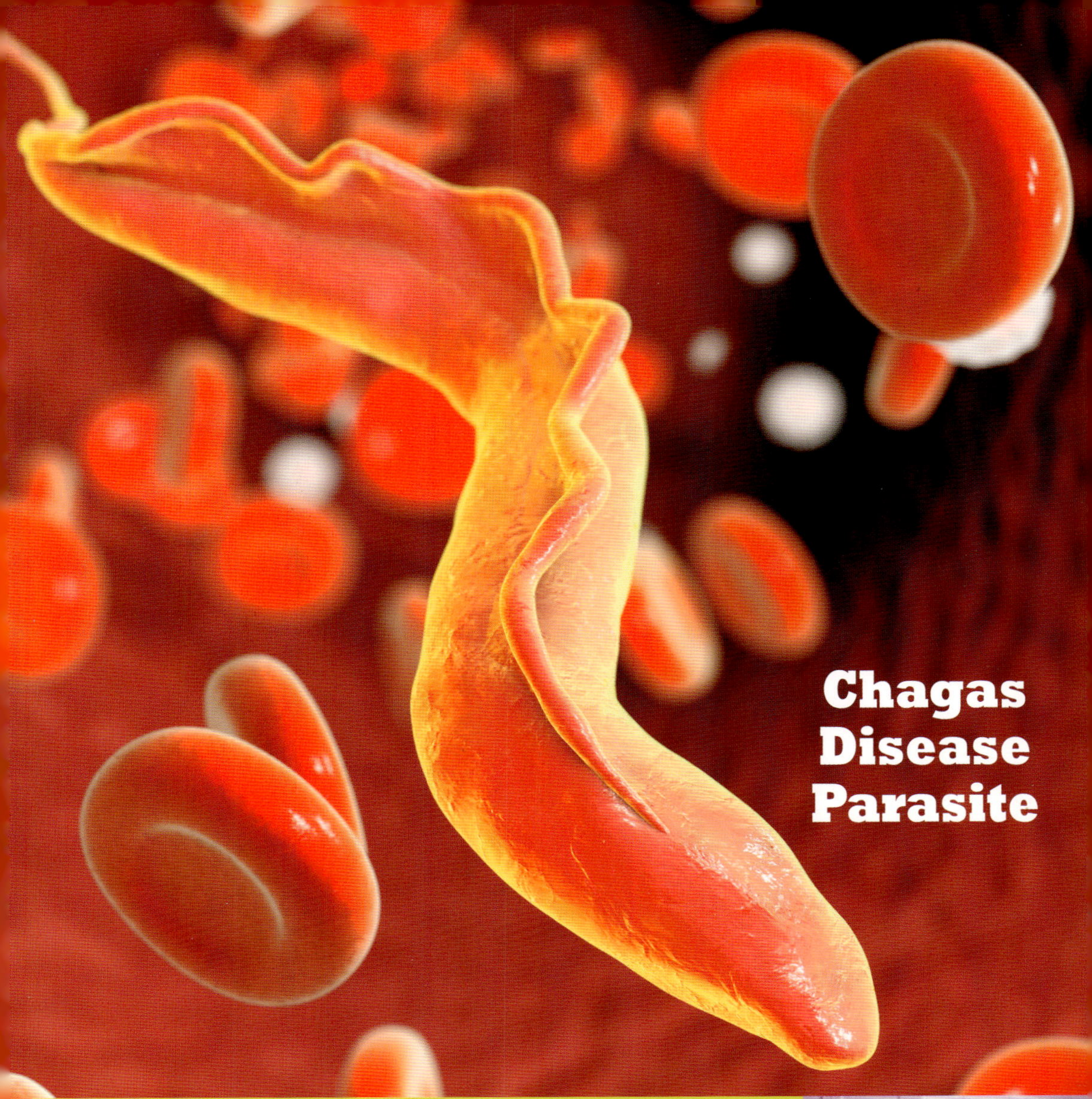

Chagas Disease Parasite

Dr. Carlos Chagas discovered the parasite in 1909. But it's been around for much longer. Scientists have found the parasite in a 9,000-year-old mummy.

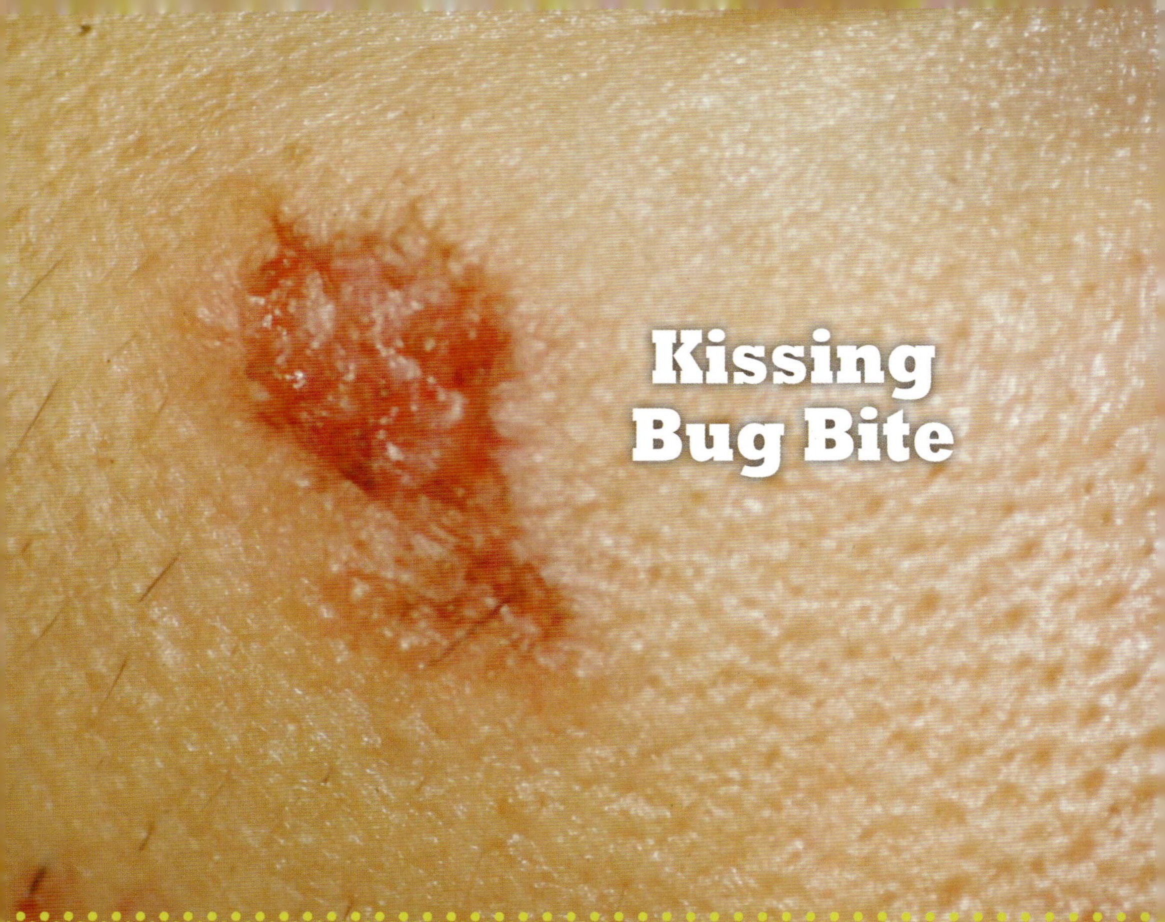

Kissing Bug Bite

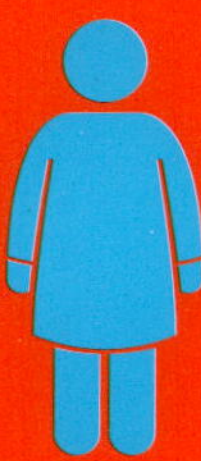

= 20 people

About 20-30 out of 100 people infected with Chagas disease will develop life-threatening problems.

Stay Away!

A kissing bug should never be touched or squashed. That might spread the parasite. The bug should be captured in a container. Any surfaces it walked on should be cleaned. Windows and doors should be sealed to keep out these dangerous bugs.

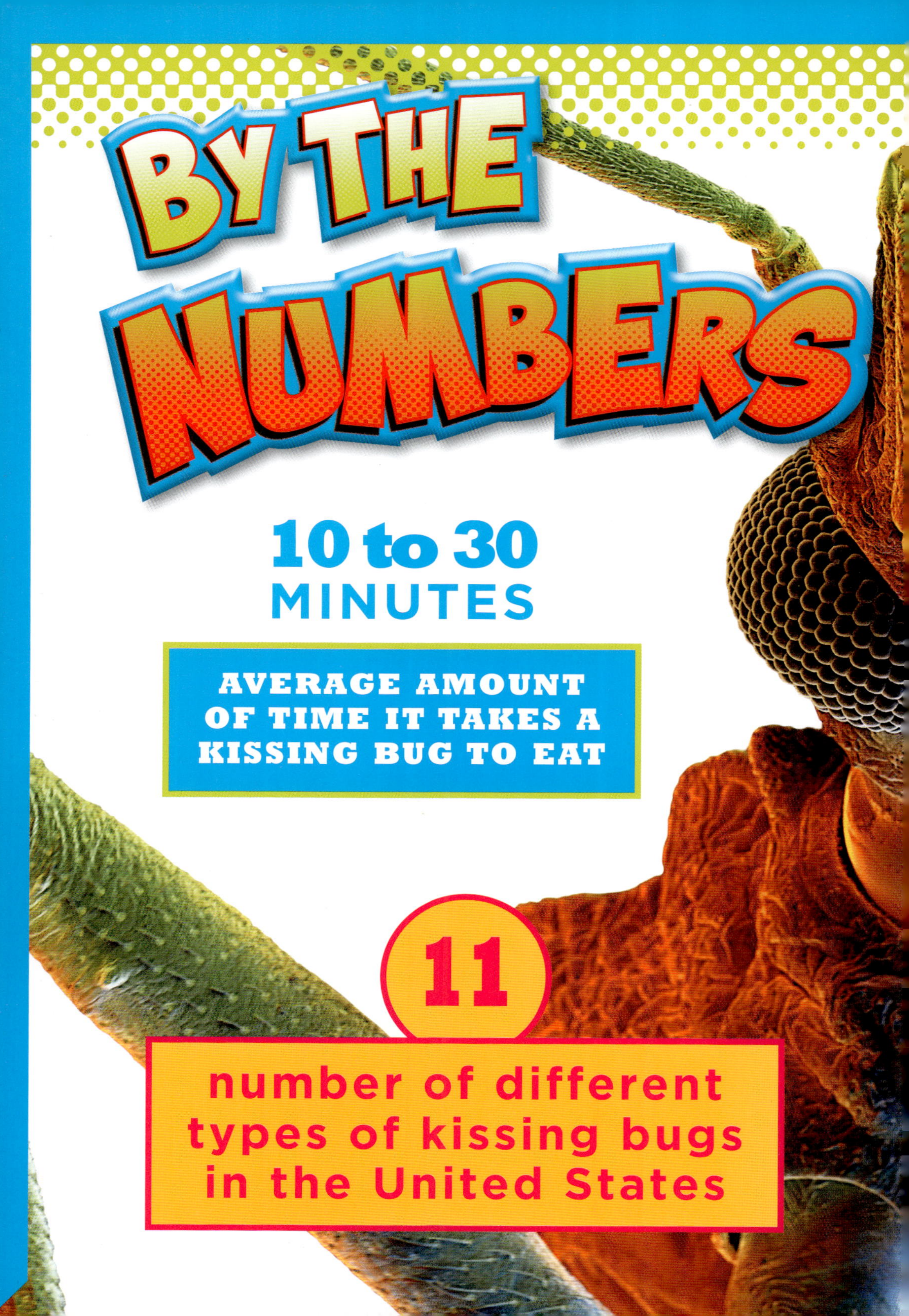

BY THE NUMBERS

10 to 30 MINUTES

AVERAGE AMOUNT OF TIME IT TAKES A KISSING BUG TO EAT

11

number of different types of kissing bugs in the United States

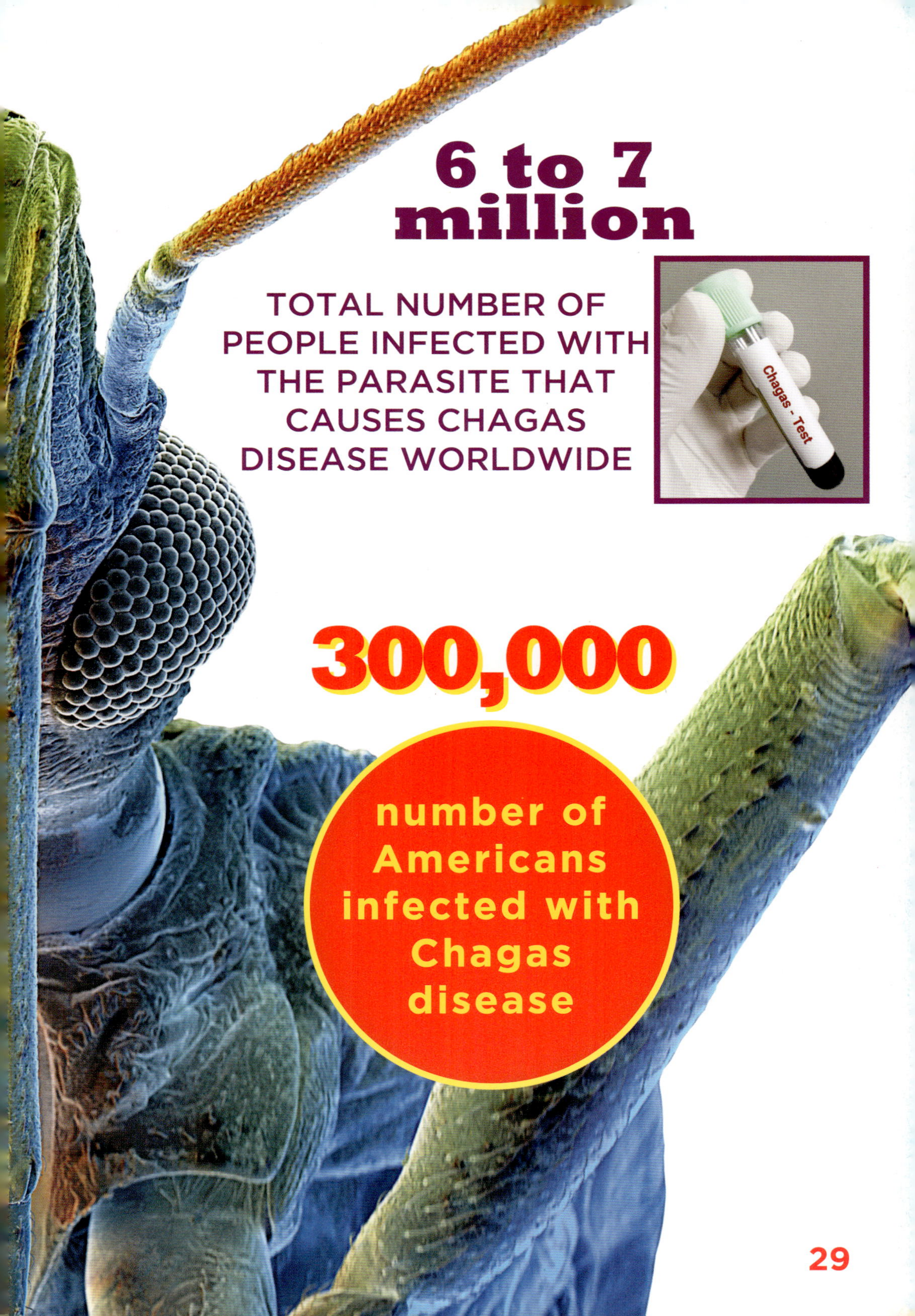

6 to 7 million

TOTAL NUMBER OF PEOPLE INFECTED WITH THE PARASITE THAT CAUSES CHAGAS DISEASE WORLDWIDE

300,000

number of Americans infected with Chagas disease

GLOSSARY

abdomen (AB-duh-muhn)—the rear part of an insect's body

appetite (AH-puh-tyt)—the desire to eat

assassin (uh-SAS-in)—someone who kills another person, usually for pay or from loyalty to a cause

burrow (BUR-oh)—a hole in the ground made by an animal for shelter or protection

exoskeleton (eks-o-SKEL-luh-tuhn)—the hard, protective cover on the outside of an insect's body

parasite (PAR-uh-syt)—a plant or animal that lives in or on another plant or animal and causes harm

pierce (PEERS)—to make a hole through

proboscis (prah-BAH-sis)—a long, thin tube that is part of a kissing bug's mouth

BOOKS

Blake, Kevin. *Killer Kissing Bugs.* Bugged Out! The World's Most Dangerous Bugs. New York: Bearport Publishing, 2019.

Hofer, Charles C. *Tiny but Deadly Critters.* Killer Nature. North Mankato, MN: Capstone Press, 2022.

Levy, Janey. *Lethal Insects.* Mother Nature Is Trying to Kill Me! New York: Gareth Stevens Publishing, 2020.

WEBSITES

Assassin Bug Facts for Kids
kids.kiddle.co/Assassin_bug

How a Kissing Bug Becomes a Balloon Full of Your Blood
https://video.idahoptv.org/video/how-a-kissing-bug-becomes-a-balloon-full-of-your-blood-vukqnu/

Kissing Bugs
www.sandiegocounty.gov/content/sdc/deh/pests/vector/kissingbug.html

INDEX